Six-Word Lessons for

THE DRONE HOBBYIST

100 Lessons to Safely Fly Your Drone with Skill and Confidence

2nd Edition

Dr. Tulinda Larsen
Ruth Blomquist

Published by Pacelli Publishing
Bellevue, Washington

Six-Word Lessons for the Drone Hobbyist

All rights reserved. No part of this book may be reproduced or transmitted in any form or by any means, electronic or mechanical including photocopying, recording or by any information storage or retrieval system, without the written permission of the publisher, except where permitted by law.

Limit of Liability: While the author and the publisher have used their best efforts in preparing this book, they make no representation or warranties with respect to accuracy or completeness of the content of this book. The advice and strategies contained herein may not be suitable for your situation. Consult with a professional when appropriate.

Copyright © 2018, 2021 (2nd Edition) by Tulinda Larsen and Ruth Blomquist

Published by Pacelli Publishing
9905 Lake Washington Blvd. NE, #D-103
Bellevue, Washington 98004
PacelliPublishing.com

Cover and interior designed by Pacelli Publishing
Cover image by Pixabay.com
Author photo by Susan Casley

ISBN-10: 1-933750-85-5
ISBN-13: 978-1-933750-85-9

Dedication

This book is dedicated to my husband, Carl Larsen, who is my drone catcher.

Testimonials

"Dr. Larsen's book, Six-Word Lessons for the Drone Hobbyist, *is contributing to the advancement of drones as a hobby and provides an excellent understanding of the fundamentals."* --Sharon Rossmark, CEO and Founder, Women And Drones, LLC

"As a leading drone training company, I found the 100 Lessons Dr. Larsen provides in Six-Word Lessons for the Drone Hobbyist *to be a great start for enthusiasts who are looking to venture into the unmanned world as a drone operator."* --Bryan McKernan, Vice President North America, Consortiq

"I am a drone hobbyist and I have been operating a drone on my farm for more than a decade. The 100 lessons Dr. Larsen provides in this book are exactly what drone hobbyists need to know to improve operational performance." --Robert Blair, Three Canyon Farms

"As a marine biologist and a drone pilot, I believe that every new skill starts with taking small steps. Learning flight skills before launching into a business is where Dr. Tulinda Larsen skillfully designed this book to help new drone pilots take their first steps in this emerging industry." --Alicia Amerson, MAS, PMP, CEO of Alimosphere

Contents

Introduction..7
Acknowledgements 8
Introduction to Exciting World of Drones 9
Finding Fun Drone Things to Do............................15
Choosing the Right Drone for You..........................33
Unpacking Box When Your Drone Arrives..............41
Understanding Basic Principles of Drone Flying.....47
Before You Take Your Drone Flying........................57
Learning to Fly Your Drone Safely..........................77
Prevent and Recover Loss of Drone 95
Finding Other Drone Enthusiasts and Groups.....109
Thinking About Starting a Drone Business? 119

Introduction

My second edition provides updates based on new FAA rules for drone operations.

After more than 40 years in aviation, including getting my private pilot license in Alaska, I have turned to flying drones. If you already own a drone or are thinking about buying a drone as a hobby, this book is for you. I have packed into 100 lessons all the things I wish I had known before I started flying my drone. All of these lessons come from my experiences learning to fly a drone and crashing two drones.

While it is true drones pretty much fly themselves, there are things you need to know to get the most fun out of your drone and to fly safely.

Acknowledgements

For this second edition, I thank Ruth Blomquist who worked with me to update the sections. Ruth came to me as a high school intern when I was Executive Director of Utah's Deseret UAS. As part of the internship, I taught her to fly drones and to ultimately pass her Part 107 Remote Pilot in Command test. She went on to form a drone club for her high school and held a drone day for Utah Girl Scouts. At the writing of this book, she is now a sophomore at Utah State University and is an active drone pilot with the USU AggieAir program. She is also part of the team for ElectraFly, the personal flying machine.

I also thank Darryl Jenkins, who changed my life by introducing me to the exciting world of drones.

Introduction to Exciting World of Drones

1

Drones are a fast-growing hobby

Drones came into the consumer market around 2010 with improvements in miniaturization technology making it possible to manufacture reasonably priced drones. In 2015, the U.S. government started requiring all drones between ½ pound and 55 pounds to be registered. By 2020, 2 million drones were registered, with 1.4 million operated by hobbyists. For perspective, it took 100 years for the fleet of manned aircraft to reach 250,000.

Understanding the definition of a drone

They go by many names--Unmanned Aerial Systems (UAS), Unmanned Aerial Vehicles (UAV), quadcopters, multi-rotor aircraft, but more commonly they are known as drones for the sounds they make when in flight. Drones are remotely piloted aircraft and come in a wide range of sizes from as small as the palm of your hand to high-flying military aircraft.

Hobby drones are fun to fly

Drones are relatively easy to learn to fly as a hobby. There are many YouTube videos that teach the basics. Drones also have intelligent flight modes that can be set to automatic modes, such as following you, flying a 360 around your house, flying a specific flight path, and taking instructions from your gestures. Images can be livestreamed or processed after the flight using consumer-friendly software.

What you'll learn from this book

Now that you are ready to jump into the exciting world of drones, these lessons will provide examples of fun things to do with your drone. You will also learn to: choose the right drone, unpack the box when it arrives, fly your new drone, know what to do if your drone crashes, find other drone enthusiasts to fly with and learn from, and gain some ideas on turning your new hobby into a business.

Finding Fun Drone Things to Do

5

Imagination only limits things to do

Drones provide a bird's-eye view of the world with breathtaking images and videos. What you can do with your drone is only limited by your imagination. This chapter will discuss several ideas for using your drone as a hobby. I would love to hear your ideas!

Drone pictures and videos are amazing

Drones not only have incredible, cutting-edge technology, but the cameras mounted on them are better quality than most high-end DSRL (digital single reflex lens) cameras. They are certainly far better than any mobile phone camera, which gives them another advantage over any other type of photography.

7

Capture unforgettable family moments and memories

Most people start with taking stunning pictures of events and special occasions, usually for friends and family. Who doesn't like to look up and wave to a drone! You can even livestream to Facebook or YouTube. Out-of-towners can be part of the action.

Special wedding day memories from above

On that special wedding day, you can provide friends and family with extraordinary pictures and videos from above. Just think about the beauty of the bride and groom from the sky with a 360-degree video of the venue. This is a gift most people cannot provide. If you sell the photos, you need a drone pilot license.

9

Get the best party pictures ever

Just imagine the pictures you could capture from above at your family reunion, a floating boat party, or a party on the beach. Once again, you can livestream to Facebook or YouTube to share the party. This allows you to include friends and family who don't live in the area and can't be part of the action.

10

Capturing nature's beauty, magnificence, and inspiration

Capturing nature is one of the most popular uses of a drone by hobbyists. When flying your drone over nature and animals in their natural habitat, it is important to be sensitive to the impacts on wildlife. Check out *Six-Word Lessons for Drone Pilots and Outdoor Enthusiasts* by Alicia Amerson for the best practices of flying over nature and wildlife.

Recording sporting events, achievements, and records

Drones are perfect for covering outdoor sporting events and achievements, maybe even capturing records. A drone can cover friends and family striving for their best performance. You will need a drone pilot license if your drone images are used to score races, sell photos, or use the photos for promotion of the event.

Covering running events start to finish

A really cool feature of many drones is called "follow me" or "track," where the drone will follow a person or other object. You can have the drone follow you in a race or track your friends. Once again, you will need a drone pilot license if your drone images are used to score races, sell photos, or use the photos for promotion of the event.

Climb the mountain for incredible pictures

As you climb a mountain, you can have a small drone in your backpack and take it out to capture the magnificence of the mountain scenery. You can use your drone to follow friends as they climb or fly ahead and check out the trail. Importantly, drones can help ensure safe climbing in many ways.

14

Capture your boat underway, sailing, fishing

Boats and drones provide some of the most amazing images. You can set the drone to follow your boat underway and circle to get a 360-degree view. You can send your drone ahead to find fish and other wildlife in the water.

Cover crew teams on the water

Watching the gracefulness of a crew team glide along the water is so beautiful. You can send your drone out to capture the starts and finishes. Of course, you will need a drone pilot license if your drone images are used to score races, sell photos, or for promotion of the event.

Capture majestic alpine winter sport pictures

You can pack a small drone in your backpack and take majestic pictures and videos, beyond what a GoPro camera can capture, when skiing or snowboarding. For back country skiing, your drone can fly ahead and look for dangerous situations, such as potential avalanches. Just make sure the area allows drone operations.

# Improving sport performance to meet goals

Whatever your sport, you can use the drone images to improve your personal performance. You can capture yourself from above, in front, behind, and on the side. Then review the videos and photos to critique yourself and improve your performance.

Planning sport activities and safety checks

You can even use your drone to plan a sporting activity and scout out the best courses. When setting up an activity such as an obstacle course, bike run, or swimming competition, your drone can check out the course and make sure it is safe.

Record kids sports for practice lessons

Parents love to take pictures of their kids' activities. Drones provide a new perspective to enjoy and share with friends and family. You can livestream to Facebook or YouTube to share with out-of-town family. As mentioned in previous lessons, you will need a drone pilot license if your drone images are used to score races, sell photos, or use the photos for promotion of the event.

20

You can even race your drone

Drone racing is a rapidly growing sport around the world. Racing drones are very small and compete based on speed and accuracy through an obstacle course, either indoors or outside. The competitors put on goggles and virtually get inside the drone with a First Person View (FPV). Check out The Drone Racing League, MultiGP, International Drone Racing Association, DR1Racing, or Rotorcross.

Taking your drone through TSA screening

From personal experience, I can tell you TSA does not have any problems with clearing consumer drones. You need to take the drone as carry-on luggage and keep the batteries separated. This will allow you to take your drone with you on big manned aircraft to really fun places.

Choosing the Right Drone for You

Price ranges for purchasing your drone

You can buy a drone for less than $100, but they are really toys. Cheap drones are not as much fun to fly because they are much more difficult to control. Most drones operated by hobbyists cost between $300 and $1,300. These consumer drones have amazing technology packed into very small aircraft. The more expensive drones have better cameras and gimbles.

Hobby drones for less than $100

There are toy drones, such as flying Tinker Bell or Ninja Turtles, that sell for less than $20. While they are fun to play with, they really are not hobby drones. Hobby drones for around $100 have cameras and some of the intelligent flight modes, plus they can do flips and other tricks. With these drones, you can enjoy a hobby drone with very little expense.

Six-Word Lessons for the Drone Hobbyist

24

Hobby drones for less than $1,000

Drones less than $1,000 are in the midrange of capabilities for hobbyists. The cameras have more features and have much better collision avoidance than the less expensive drones. They are also lightweight and have less wind tolerance than the more expensive drones.

25

Hobby drones for more than $1,000

The more expensive drones have better wind tolerance and much better cameras that are mounted on gimbles to get incredible shots. In addition, you can change the cameras on these drones for different photo shoots. As the drones get more expensive, the ability to automatically avoid collisions improves.

Best drones to win drone races

A racing drone is a small quadcopter that is built to compete in FPV (first person view) racing events held in most major cities around the world. Racing drones are not the same as camera drones discussed above. Camera drones fly "slow and low" and are designed to capture high-quality video. A racing drones' camera is designed for speed. Winning drones are built, not bought.

Deciding to purchase your first drone

When deciding to buy a drone for a hobby, first figure out your budget. Then choose the drone that best matches the drone activities you are considering. Everyone crashes or loses a drone, so you need to budget for repairs or replacement. More expensive drones have replacement insurance plans.

Where can you purchase your drone?

Hobby drones can be either impulse purchases or well-planned. Drones can be purchased online through the manufacturer websites or other websites and stores, such as electronics stores. Toy drones can be purchased almost anywhere.

Unpacking Box When Your Drone Arrives

Check the contents of the box

When you open the box, first check the contents against your order. Take out the drone, remote controller and other pieces. You will need to charge the batteries and remote controller. It is always best to follow the manufacturer's instructions to assemble the drone.

Decide how to transport your drone

You will need something to store and transport your drone with the accessories. There are a lot of options including backpacks, soft cases, hard cases, and cases on wheels. It is important that the case has individual sections to safely store extra batteries.

31

Accessories might not be in box

In addition to the drone, propellers, and remote controller, there are other things you need to operate your drone, such as a mobile phone or tablet to insert into the remote controller. Extra batteries, SD cards, and propellers are good ideas. A multiple battery charger is another good accessory to keep everything topped up.

Register and get ID from FAA

All drones weighing between ½ pound and 55 pounds, including the payload, must be registered. Go to faaregisterdrone.com and follow the instructions. DO NOT get sucked into sites that charge fees to file the paperwork. Registering your drone is very easy.

How to make your drone legal

Registration costs $5 per drone and is valid for 3 years. To register, you'll need an email address; credit or debit card for the $5; physical address and mailing address (if different from physical address); and drone make, model and serial number. You must be 13 years or older to register a drone, so anyone younger than 13 needs to have someone older register the drone. You must also be a U.S. citizen or legal permanent resident.

Understanding Basic Principles of Drone Flying

What goes up must come down

As a drone hobbyist, it is important to understand the principles of how a drone flies and how to bring your drone back to earth safely. Drones are different from remote control helicopters because they can operate with a level of autonomy through preprogrammed functions.

Importance of balanced positions for propellers

Spinning propellers push air down. All forces come in pairs, which means that as the rotor pushes down on the air, the air pushes up on the rotor. This is the basic idea behind lift, creating the control of the upward and downward forces. The rotors spin two of the propellers clockwise and two counter-clockwise, which controls the direction of flight. It is important to install propellers exactly as the manufacturer recommends.

36

Communication between remote controller and drone

The remote controller communicates with the drone using radio waves. Drones are typically run on 2.4 gigahertz radio waves. To communicate with the drone, the remote controllers use Wi-Fi, which can be transmitted on the 2.4 gigahertz spectrum, and is something that smartphones and tablets can tap into without any accessories.

Understanding LiPo batteries good and bad

LiPo batteries power all of the electronics and motors on your drone. LiPo batteries are based on Lithium Polymer chemistry (hence the name LiPo) making them lightweight with a lot of power. But these batteries only have charge of about 30 minutes and are very sensitive to extreme temperatures. I never leave my drone in the car because of the LiPo batteries' sensitivity to temperature.

Understanding GPS for drone position locking

A GPS chip inside the drone relays its location to the controller and helps to keep the drone on track. It also logs the aircraft's takeoff spot in case it needs to "Return-to-Home" unassisted. You can fly a drone without GPS, but GPS makes the drone much easier to fly.

Gyro-stabilization essential for keeping drone steady

Gyro-stabilization technology and Inertial Measurement Unit (IMU) make a drone fly smoothly even in strong winds and gusts. An IMU works by detecting the current rate of acceleration using one or more accelerometers. The IMU detects changes in rotational attributes like pitch, roll and yaw using one or more gyroscopes. Drone smooth flight capabilities allow the filming of fantastic aerial views of friends, family, and nature.

Basic remote controller functions and operations

The remote controller controls the drone using two joy sticks and several special buttons. The joy sticks control the drone's movements, while the buttons can control the camera gimble, Return-to-Home, and the flight modes discussed in Lesson 74. A mobile device can be connected to the remote controller to monitor the drone's telemetry, view the images being captured, and adjust the camera settings.

Preventing crashes with collision avoidance systems

Drones with obstacle detection and collision avoidance sensors keep the drone from crashing into things. But the collision avoidance technology cannot sense thin objects, such as power lines and tree branches, nor can it sense things coming directly at the drone.

Before You Take Your Drone Flying

FAA regulates drone flying for hobbyists

The FAA has specific regulations for drone hobbyists, stated as: "flying for enjoyment, recreation, outside of work and not for work, business purposes, or for compensation or hire." These drones are considered model aircraft and FAA does not require a drone pilot license or insurance.

Fly under special model aircraft rule

FAA has a free, online test designed to provide recreational drone pilots with "aeronautical safety knowledge and an overview of the rules for operating drones in the National Airspace System." Pilots are required to follow community-based safety guidelines and fly within the programming of a nationwide community-based organization, such as the Academy of Model Aeronautics discussed in Lesson 90.

The rules for flying over people

If you are a hobbyist, you may not fly your drone directly over people unless the pilot has their permission to fly over them or they are under something that offers enough protection.

Check restrictions to fly near airports

Hobbyists must notify airports within 5 miles of where they plan to fly. However, hobbyists are not allowed to operate in "controlled airspace" at commercial airports. There are some great apps, including the FAA's B4UFLY app, to get information about the airspace around airports and airport contact information.

Never fly drones near other aircraft

Drones must always give right-of-way to manned aircraft to avoid collisions. It is much easier for a drone pilot to see a manned aircraft than for a manned aircraft pilot to see a drone. Manned aircraft include big jets and smaller general aviation aircraft, helicopters, parachute jumpers, and even hot air balloons.

Always keep your drone in sight

The FAA requires that you keep your drone within visual sight, approximately .8 miles. You cannot use binoculars to follow your drone. A Visual Observer is good to watch the drone and look for any potential hazards while you operate the drone. There are special regs for night flying, including anti-collision lights that can be seen for 3 sm and have a flash rate that is sufficient to avoid collision.

Never fly higher than 400 feet

The FAA set the maximum altitude for drones at 400 feet above ground level because general aviation aircraft are at 500 feet and above. Most hobbyist drones are geofenced to not climb higher than 400 feet. It is important to continually look for other aircraft at all altitudes, especially helicopters.

Never fly near emergency response efforts

Drones operated by hobbyists are very disruptive to emergency response teams at times of natural and other disasters and are not allowed to be flown over disaster sites. Commercially operated drones may provide assistance in times of emergency under special exemptions.

Check restrictions to fly national parks

The National Park Service has made it illegal to launch, land, or operate drones. But you should check with the park service. You might get permission to fly in certain areas. The FAA has restricted drone flights above 10 national monuments. The FAA B4UFly app and the free app AirMap can provide the information you need to make sure your flight is legal.

Check local community restrictions to fly

Communities are responding to privacy complaints from citizens and are passing rules restricting where drones can be operated. They also see an opportunity to make money with registration fees. For the most part, the FAA has the authority over drone operations while local communities do not. However, local communities can restrict take-offs and landings in parks, but cannot restrict flying over a park, because the FAA controls airspace.

52

Prepare for possible flight emergency situations

Emergencies do happen. Think through possible emergencies and prepare your responses before the worst happens. For example, what if you lose contact with your drone, what if you lose sight of your drone, or what if your drone crashes?

Make a checklist and follow it

Use a checklist to make sure everything is working properly before each flight. Check such things as: batteries are full and in good condition before inserting into drone; propellers don't have any nicks and are installed properly; the SD card is inserted; the connection between your drone and the remote controller is correct.

Check your flight area for obstacles

Before launching your drone, go out and inspect the site for any potential obstacles that your drone's collision avoidance might not see. These include power lines, towers, and trees. You can use your drone to look for obstacles. Send your drone to a safe height and then using the camera, do a 360 view to see if there are any obstacles.

Be considerate of invading personal privacy

A major issue for drones is invasion of privacy by hobbyists flying over and taking pictures of unsuspecting folks. You can be almost anywhere and a drone can pop over the horizon and fly near you with its camera looking straight at you. There is nothing people can do to stop the invasion of privacy, since drones are aircraft and it is a felony to shoot or disable aircraft. The only action is to call local law enforcement and have privacy laws enforced.

No insurance required but good idea

Insurance to operate a drone as a hobbyist is not required by law. Some manufacturers offer replacement drone insurance if you lose your drone. Flying without coverage exposes you to significant lawsuits. Some homeowner insurance policies include drones, so you should call your agent to find out specifically what your policy does and does not cover.

Put a label on your drone

The FAA drone registration number must be marked on your drone. I recommend making a sticker for the top of your drone with the registration number and your contact information in case the drone gets lost. I might have gotten my first lost drone back if I had done that.

58

Make sure you have enough storage

Files from the drone's camera on the SD card can be as large as 3 to 5 gigabytes for a 15-minute video. Make sure you have enough hard drive capacity to store the files. It is so frustrating to take a large number of pictures and videos then not be able to load them on your computer to edit or store them.

You will need image editing software

Some drones come with very simple video editing right in the control software. To do really cool things with your drone videos, like adding special effects and music, you need video editing software. iMovie or FinalCut are good for Macs. Adobe Premier Pro CC and Corel VideoStudio work for PCs.

Learning to Fly Your Drone Safely

60

Watch and learn from YouTube videos

While you are waiting for your drone to arrive, go to the manufacturer's website and YouTube and watch videos on how to operate your drone. You can also check out Facebook groups to get great ideas on things you can do with your drone. It is amazing how much you can learn from others!

Check flight restrictions on drone apps

Install the FAA B4UFly app onto your mobile device to make sure you know flight restrictions in the area where you plan to fly. There are flight-planning software products that provide air space information that can help you plan your project.

62

Check weather status on drone apps

Weather dramatically impacts the performance of drones. In my opinion the best weather app is UAV Forecast. It provides a very simple Go/No Go display, sunrise and sunset times, winds and temperature, possibility of precipitation, and the number of satellites available.

63

Start by flying in beginner mode

Beginner mode will ensure that the drone does not fly away farther than 100 feet horizontally and vertically. Also, the speed is reduced about 50 percent. By restricting the performance of your drone, fewer bad things can happen at slower speeds.

Practice basic maneuvers for drone flights

The drone is controlled by little joy sticks on the remote controller. Watch manufacturer and YouTube videos to learn how to use the joy sticks to fly your drone. Remember, practice with the drone's camera facing out. If you turn the drone around, all the commands with be reversed.

Practice flying drone up and down

The first skill you need to master is take-off and landing. To take off manually, push the left joy stick straight up. Some drones have an automatic landing feature, or you can land the drone manually by pulling down on the left joy stick. Practice take-off, hover and landing until you are comfortable. Practice for at least one battery charge.

66

Practice flying drone left and right

Once you are competent to take off, hover and land, you should practice, turning the drone around on its own axis by using the left joy stick and pressing to the left for clockwise rotation and right for counter clockwise rotation. Next, practice flying left and right by using the right joy stick and pushing to the left and right. Again, practice for at least one battery charge.

Practice having your drone return home

Everything is going great, but all of a sudden you lose control and orientation of the drone, so you can hardly see it. Many drones have a helpful one-key feature called Return-to-Home (RTH). You need to set RTH to a specific place or to return to the location of the remote controller. Take the time to practice RTH so you can activate it quickly.

Practice advanced flight operations: Follow Me

Follow Me and ActiveTrack follow the object or person. The object being tracked needs to contrast with the background. The flight altitude can be changed at any time. You can also command the drone to automatically circle the subject, fly backwards in front of the object, fly behind the object or fly alongside the object.

Practice advanced flight operations: 360 Circle

This feature is known as Point-of-Interest. This intelligent flight mode allows you to set the center on something, like a house or boat, and then set the diameter of the circle and adjust the altitude. Once you hit go, the drone completes a perfect circle around the object. Be sure to account for any obstacles.

Practice advanced flight operations: Tripod feature

Tripod mode slows things down and makes the joy sticks slower to respond to achieve truly professional cinematic shots, as if the camera were on a dolly. It is great for flying in tight spaces, because the drone is easier to control.

Practice advanced flight operations: Waypoints feature

Waypoints and Draw features can be used to expand your flight capabilities and enable you do more with your drone. Simply draw a route on-screen and your drone will move along the path while keeping its altitude locked. This lets you focus on camera control and enables more complex shots.

Practice advanced flight operations: Gesture Mode

Using Gesture Mode, selfies can be captured easily using a few gestures without the remote controller. You simply lift your arms when facing the camera and the drone will recognize this movement by locking on and placing you in the center of the frame. A 3-second countdown will begin, giving you time to strike a pose, allowing moments to be captured without the remote control in the picture.

Practice advanced flight operations: Terrain Follow

Terrain Follow mode is used to maintain a height above ground between 3 and 30 feet. This mode is designed for use on grassland sloped at no more than 20 degrees. The drone will maintain the recorded height during flight and ascend as the slope rises. However, the drone will not descend in downward slopes.

Practice Position, ATTI and Sport Modes

Position or P-Mode is the default mode of your drone and activates GPS positioning. Altitude, or ATTI Mode, switches off the GPS. This mode is for experienced pilots, allowing them to fly very steady and smooth, which can be great for filming, but you don't have RTH. In Sport, or S-Mode, your drone flies like a hotrod at speeds up to 45 miles per hour. This is good in high winds when you need more power.

Special considerations when flying over water

Reflections off the water can sometimes confuse the positioning sensing of your drone. Be sure to stay away from obstacles that could interfere with your drone's compass or signal, such as lighthouses, boat transceivers, antennas, and other electronic devices. Don't fly your drone near birds because they may attack it.

Finding a place to practice flying

Finding a place to practice flying your drone is sometimes difficult. You are not allowed to fly on most school fields, unless you get permission. Parks may have local restrictions. Some parking lots are good practice spots. If you join Academy of Model Aeronautics (AMA) as explained in Lesson 90, you can practice on AMA fields.

Prevent and Recover Loss of Drone

Never fun to lose your drone

Losing a drone is one of the worst things that can happen when flying your drone. Due to the nature of drones, you can see a $1,000+ investment fly away in a matter of minutes, never to be seen again. In this chapter I will discuss the reasons your drone might fly away and what to do if your drone crashes.

Problem: not knowing your drone's orientation

A major reason you can lose your drone to flyaways is if you lose orientation of your drone. Once the drone is far enough away, many novice users struggle to figure out which way it is facing and to return its orientation to come back home. This is the reason you should practice flying in beginner mode when operating more expensive drones.

79

Problem: lost connection with drone flying

Lost connections can also cause you to lose your drone. You should be aware of the range of your drone's connection with the remote controller. By flying your drone beyond its range and losing connection you will have a very hard time regaining control of your drone again to perform a safe landing.

Problem: signal interference impacting drone performance

Interference is another common reason you could lose your drone. You should be aware of the frequency of your drone and any potential interference in the area you are flying. Interference can cause a loss of control of the drone and cause it to perform erratic actions, possibly leading it to either flyaway or crash.

Problem: drone reaches critical battery level

First off, make sure you always monitor your drone's battery life. The best drones on the market have a battery life of around 25 to 30 minutes. This isn't that long, especially considering how time flies when you're having fun. Make sure you know how long your model can fly and don't let the battery percentage hit the single digits, or you may find yourself in trouble like I did when I lost my first drone.

Problem: head wind drains battery rapidly

It is very important to know the wind speed, especially if it is a windy day, because flying against a headwind takes a lot of energy. Monitor your battery levels to be sure you have enough battery charge left to fly back.

Problem: drone falls out of sky

Drones can fall out of the sky if they hit something, such as power lines and tree branches that can't be detected by collision avoidance. Also, some modes turn off collision avoidance, allowing your drone to fly into things.

Problem: someone shoots down your drone

Shooting down a drone is a federal offense. Federal statute also prohibits interfering with anyone "engaged in the authorized operation of such aircraft" and carries a penalty of up to 20 years in prison. Since drones are considered aircraft, threatening a drone or a drone operator is also a federal crime subject to 5 years in prison under this same statute.

Find out when to notify government

Neither the FAA nor the National Transportation Safety Board (NTSB) require accidents involving drones that are flown for hobby or recreational use to be reported. NTSB stated that it has "consistently excluded UAS flown for hobby and recreational use from its accident rule and has historically not investigated the rare occasions in which a model aircraft has caused serious injury or fatality."

Requirements if drone causes human injuries

Surprisingly, there is no requirement for hobbyists to report any injuries if your drone causes injuries to someone. Frankly, the propellers are designed to break away on impact and tests have shown that only minor injuries occur when drones hit people.

Find every part if drone crashes

Realistically, you will crash your drone. As I said in the introduction, I have lost two drones. If your drone crashes, try to retrieve the drone and gather up all the parts, including propellers, batteries and any other parts.

88

Insurance might provide a replacement drone

For a small insurance premium, some manufacturers will replace your drone in the first year if you can return all the damaged parts. The manufacturer may repair your drone if the damage is not structural. The Aircraft Owners and Pilots Association (AOPA), discussed in Lesson 92, offers insurance that includes payment of up to $2,000 for drone damage.

Finding Other Drone Enthusiasts and Groups

89

Find other hobbyists to improve skills

The best way to learn how to be a better drone pilot is to fly with others! While there is no centralized way to find other drone hobbyists, consider these: the Academy of Model Aeronautics (AMA), Association for Unmanned Vehicle Systems International (AUVSI), Facebook groups, Drone User Group Network (DUGN), Aircraft Owners and Pilots Association (AOPA), drone racing groups, Women and Drones, and local Meetup groups.

Academy of Model Aeronautics has benefits

The Academy of Model Aeronautics (AMA) (ModelAircraft.org) includes drone hobbyist in the membership. A big benefit for drone hobbyist is the AMA insurance as mentioned in Lesson 56. AMA also has sanctioned aerodromes, which can be a good place to practice as mentioned in Lesson 76.

91

AUVSI gives guidance to drone hobbyists

The Association for Unmanned Vehicle Systems (AUVSI.org) is the largest organization for drone operators, manufacturers, and service providers. AUVSI's focus is on advocacy for the drone industry before government agencies. But as the drone hobbyists grow in numbers, AUVSI's local chapters are open to drone hobbyists.

AOPA has dedicated Drone Pilot Group

The Aircraft Owners and Pilots Association (AOPA.org) represents private pilots and has a Drone Pilot membership. AOPA provides helpful services for drone hobbyists including: Drone Pilot e-newsletter with up-to-date information; AOPA Pilot Information Center toll-free hotline to answer questions about drones; safety and education; events at AOPA Fly-Ins; and YouTube livestreams and podcasts on the latest drone news.

Use Facebook groups for sharing information

There are numerous Facebook groups that bring together drone hobbyists. Drone racers have a great group called Drone Racing International FPV. Drone hobbyists can learn about best practices to be environmentally sensitive on *The Social Innovation Drone Tribe*. Other groups include *Droners* by squadDrone, focused on sharing travel experiences, and *Drone Gossip*, which shares news and other information.

94

Drone racing groups are very popular

As mentioned in Lesson 20, there are several drone racing groups. These groups have a lot of participants and provide blogs and other information on the exhilarating sport of drone racing. As an added benefit, you will meet other drone racers at the events.

Try Women and Drones for information

Women interested in drones as a hobby can go to Women and Drones (WomenandDrones.com) which encourages women to learn to fly drones and features really interesting stories about women drone pilots, like me! I was profiled in the *We Are All Awesome* project.

Check out drone publications and e-newsletters

There are many drone publications, blogs, e-newsletters, and YouTube channels. I really like *The Drone Girl* blog (TheDroneGirl.com) because she writes product reviews, tests out the latest technologies, interviews people about interesting drone products, and posts fun things to do with drones.

Thinking About Starting a Drone Business?

Operating a drone as a business

As you get really good at flying your drone and processing the pictures, you may think about starting a drone business to take pictures of real estate, weddings, sporting events or other activities and events. Lessons 98, 99 and 100 give some helpful things to think about.

98

Very different regulations and compliance requirements

The FAA defines commercial drone operations as, "flying for work, business, non-recreational reasons, or commercial gain." This typically includes flying a drone for hire, compensation, to provide a service, or for economic benefit of an entity or person. Intended use, not compensation, is the determining factor.

99

Part 107 Remote Pilot in Command

Drones operated commercially are considered FAA regulated aircraft operations under Part 107 of the FAA regulations and require a licensed drone pilot, which is earned by passing the FAA Remote Pilot Knowledge Test and completing the FAA training. In addition to passing the test, the pilot must be 16 years of age and must pass TSA security vetting. If a drone is being flown for fun, no licensing is required.

100

High penalties if you're not licensed

The penalty for operating your drone commercially, without a Part 107 Remote Pilot License, is $1,100 per violation for the pilot and $11,000 for the organization. A single flight could have multiple violations. The test is not that difficult to pass and there are several online prep courses. So if you are thinking of charging for your services, or giving your pictures to an organization for their marketing, get your license.

Contact Tulinda

For more information, or to share ideas, please contact:

Dr. Tulinda Larsen
CEO & Founder, Skylark Drone Research
TulindaLarsen.com
Tulinda.larsen@TulindaLarsen.com
Twitter: @TulindaLarsen

About the Six-Word Lessons Series

Legend has it that Ernest Hemingway was challenged to write a story using only six words. He responded with the story, "For sale: baby shoes, never worn." The story tickles the imagination. Why were the shoes never worn? The answers are left up to the reader's imagination.

This style of writing has a number of aliases: postcard fiction, flash fiction, and micro fiction. Lonnie Pacelli was introduced to this concept in 2009 by a friend, and started thinking about how this extreme brevity could apply to today's communication culture of text messages, tweets and Facebook posts. He wrote the first book, *Six-Word Lessons for Project Managers*, then started helping other authors write and publish their own books in the series.

The books all have six-word chapters with six-word lesson titles, each followed by a one-page description. They can be written by entrepreneurs who want to promote their businesses, or anyone with a message to share.

See the entire **Six-Word Lessons Series** at **6wordlessons.com**

www.ingramcontent.com/pod-product-compliance
Lightning Source LLC
Chambersburg PA
CBHW062008070426
42451CB00008BA/291